[redacted]

[redacted]

TOWARDS A TECHNOLOGY OF PEACE

[redacted]

[redacted]

TOWARDS A TECHNOLOGY OF PEACE

Theo. F. Lentz
Director, St. Louis Peace Research Laboratory

The cover design is by St. Louis artist, Fred Dreher, including a photograph of his sculpture, symbolizing the Lentz International Peace Research Award, established in 1972. He is internationally known for his work with prismatic plexiglas and other art forms.

Back cover sketch of the author is by Janice Hall.

Printed by Bruce Burgess Printing Company.

Published by Peace Research Laboratory
6251 San Bonita, St. Louis, Mo. 63105 U.S.A.

To all children with a desire to live,
born and yet to be born,
world-wide and time-wide

PREFACE

This treatise is written for those who wish to consider seriously the search for a road to a warless world. Its primary purpose, therefore, is to encourage the development of a technological attitude towards the all important problem of achieving peace.

The science of peace has moved at a less than optimum pace; it is still, however, ahead of peace technology. There is a crucial interaction between science and technology in those areas where both may be applied, but that interaction will not become operative in the enterprise of peace until the technological has been carefully developed.

Since the publication of our first book, *Towards a Science of Peace,* in 1955, the world situation has grown worse. Despite a gratifying development in the peace research movement, weaponeering has advanced so menacingly that many feel there is no hope for a net gain in the movement towards the elimination of war--that the odds are irreversibly imbalanced in the race between compassion and cupidity.

Scanning the text of our first book, we note the use of the term "peace knowhow," to which a number of our readers have taken exception. In this treatise, therefore, we take up the challenge of making peace knowhow more explicit through the concept of peace technology, examining the what, the why and the how of the matter.

The present book is a sequel to the first--an extension from peace science to peace technology. It is aimed at expediting a shift in the thinking of the individual reader from non-functioning peace wish to pragmatic peace will, and may in fact motivate some readers to examine or review the premises of *Towards a Science of Peace.*

We do not consider this text complete. It has become apparent in developing it that, ideally speaking, it should not have been published for another five years. The germinal idea would by then have been reinforced by more interviews, more dialogue, better techniques of intercommunication. But time is crucial and delay imprudent. Responses to the invitation at the end of the text will help to compensate--and to strengthen the concepts.

Our acknowledgments are numerous--space permits only the mention of a few. The author is especially grateful to Jane Abernathy, Enola Lentz and Louise Robison whose cooperation has functioned throughout the formulation of the manuscript. We are indebted to Lewis Mumford, particularly to his *Technics and Civilization,* as well as many others whose names appear in the references and in our limited bibliography. We are indebted to Mary Williams who gave us a large block of time on the analysis of the first draft, and further, to Patrick Deighan of Lancaster, England, and William Eckhardt of the Canadian Peace Research Institute. The inspiration of the latter, initiated and promoted by Norman Alcock, has been tremendous. Constant encouragement has come from our sponsoring organization, the Character Research Association--both from its contributors and from the leadership of Walter Morgan and Arun Mitra.

May 1972 — Theodore F. Lentz

CONTENTS

1. Introduction

Some thousand years ago, a Danish king called Canute II, whose strength had been proved by conquest and whose wisdom was already embodied in the reasonable and workable laws of Denmark, England, and Sweden, arrived at a logistic crisis. He thereupon ordered the sea to fall back, but the sea disobeyed. Canute, unusually introspective for a medieval man of power, considered pensively the emptiness of kingly authority unable to successfully command the sea. Philosophically, Canute's point is well-taken, but in our day not only the seas, but earth and heaven as well, have yielded at least partially to the demand of man for control. A host of land and space explorers--Galileo, Columbus, Picard, the Wright brothers, Cousteau--have challenged and penetrated the hitherto unpenetrable universe. Their success, however, is rooted in the intellectual output and inventive genius of thousands who, lacking royal expectation of obedience, put themselves to the exacting and frequently exhausting task of bringing the whole business off.

Some portion of humanity has always been eager to say, "Let there be peace!" Their earnest intent is undisputed but, like the efforts of Canute, it has had negligible effect on the waves and tradewinds of human behavior and misbehavior. Perhaps success will have to come from the work of those who have faith in man's ability to explore and resolve the enigmas of his social environment.

The peace movement is no infant; neither has it reached maturity. The individual who is inclined to join his energies to the movement finds the panorama fragmented and confusing. Organizations, certain enough of their goal, fall into a babel of discord over the most effective method of reaching it. Research centers engage in an impressive number of studies and projects; many are effectively done, most neglect the core question of how we *get* what we are after. Conferences and diplomatic enclaves produce discussions far more rhetorical than pragmatic. Indeed, rhetoric seems to be a pitfall of the peace-oriented quite as much as it is a special aid to political and diplomatic skullduggery. One may, of course, conscript the skillful phrase to one's cause--may do it in good faith and with desired effect--but rhetoric should not go untested.

The kind of sketch we have just done of obvious weaknesses does not adequately diagnose the erratic maturation process of the peace movement. Nor is diagnostic criticism as an end in itself our goal; we must, however, look sharply to the infirmities if we are to propose treatment.

A survey of the studies listed in *Peace Research Abstracts Journal,*[1] some 750 each month, indicates that for the most part, research focuses on tensions, arms control, conflicts, international relations, foreign and national policies, and military technologies; all of these have indisputable bearing on peace. Few move in a truly fresh direction. We know quite a lot about the sampling of public opinion--not nearly enough about how to utilize it. We can identify needed changes in institutional structures and value systems but have not yet delineated alternatives. Against the enormity of the task in these areas, our limited effort stands out in painful relief.

We need objective criticism of the conduct of foreign affairs, the military bastion, and industrial networks, but harangue is heady business and we must guard against excessive clatter. There is a point, perhaps subtle, at which we must look to what we humans ought to be doing, rather than at what we ought not. As we shift from negative concern with the causes of war to positive concern with the causes of peace, there should be a resurgence of energies, a release of fresh impetus which mere negativism sometimes serves to check.

Diplomacy should in fact be the principal effector of global peace but seems more often to serve it least. Disarmament conferences inspire more cynicism than confidence, peace talks breed pessimism, and national interest takes on the role of chief negotiator. When one looks at these conferences critically, he can see clearly the lack of a focus on peace as the primary goal, and only a vastly expanded conception of the interests of individual nations can overcome this serious flaw. It is difficult, if not impossible, to conceive of genuine high-level cooperation in the absence of a common cause, but as yet no one has succeeded in cutting the Gordian knot which makes boon companions of diplomacy and war.

The peace research movement was born in the era following the Second World War, when national postures had relaxed little and armament development continued apace. It is a matter of particular note that the movement survived so threatened an infancy. There are now some 40 centers or institutes throughout the world. Progress has been good, but not good enough. The work of the small band of peace researchers has not yet made it possible to negotiate the institutional and psychological obstacles which prevent the building of the peace system.

Part of the problem is quantitative. There are not enough people at work. This is true not only in research but in every facet of the peace movement. Manpower is, in fact, probably the most crucial factor in the technology herein proposed and must have highest priority. It may well be that as the number of researchers grows larger, qualitative improvements will occur simultaneously; however, it is legitimate to look at qualitative weaknesses now. Researchers in the past decade have worked at a high level of scholarship, but there is reasonable question of how to utilize their findings. Anatol Rapoport discusses this in his article, "Is Peace Research Applicable?", in the *Journal of Conflict Resolution.*[2] He points out that there is a dearth of new institutions for the application of peace research findings. With a technological orientation, peace research should be able to more precisely identify knowledge gaps and institutional requirements, to develop fresh, relevant information, and to extend and disseminate peace knowledge with a view to applicability.

Why, in the face of the current barrage of invective against technology, do we in fact propose a "technology of peace?" Widespread ambivalence towards technology is indeed justified. Not only is there personal creative attrition and environmental pollution, but the peoples of the world are sharply and tragically divided into the haves and have-nots, a cleavage in affluence that technological development has served to widen. It is a subject that seems to lend itself to heightened passions; otherwise reasonable men take equally extreme postures of attack or defense. There is no need to review their positions here; we should only be reiterating much of what we have heard time and again in the last few years.

We shall insist that only the human agent in control of the technology can be held accountable for its effects. Rather than wasting time on incessant debate, let us take a look at some suggested directions beginning with control, for as Toffler says, "The horrifying truth is that, so far as much technology is concerned, no one is in charge."[3]

In a society confronted with new proposals every day, we have no agency charged with the task of weighing technology's promises against its perils. To be sure, such an agency would have to guard against stifling innovation, but it is unrealistic to think that we can continue to avoid preventative value judgments. The most significant outcome of the U.S. Senate debate which decided the fate of the supersonic transport program may well be the debate itself; the acknowledgment of the fact that it is vital to political responsibility that advanced societies make political decisions on crucial technological questions.

"What humanity needs is not a wholesale discarding of advanced technologies, but a sifting, indeed a further development of technologies...."[4] A clear voice in the outcry says that a process of de-development and semi-development would "hasten an adequate standard of living in the underdeveloped countries and an ecologically rational standard in the developed countries...[which] would involve the abandonment of some technologies, the modification of others, and the invention of some new ones. Few would deny, we suspect, that the sooner we abandon the technologies of war....the better."[5]

We would substitute the technology of peace for that of war. If our responsibility is to separate creative potential from the capacity to destroy, then let us get at it. Technology is an engine of change, but an engine is run by human hands. If not, if it is automated, the automation is nevertheless programmed by a human mind with human sensibilities.

Peace technology can change passive desire for peace into an operational goal, and in the complex circuitry of motivation, goals and morale link closely. Three thousand years ago a poet observed that "hope long deferred maketh the heart sick" and in these still troubled times, another poet muses upon the fate of a dream deferred, left--like a raisin--to shrivel or explode in the indiscriminate rays of the sun.[6] The idealist, the pragmatist, the ripened-by-time, and the young need to believe that the goal is possible.

Face-to-face with the specter of environmental collapse, we have finally come to recognize that our earth is a closed system. We are putting values into perspective that we had brushed aside--the mystique of wilderness, the implicit wisdom of conservation, the entwined destiny of everything that lives. We live in what is frequently called a spaceship culture and spokesmen for all disciplines attest to our social interdependence. Their words form a litany to the eloquence of John Muir who said, three-quarters of a century ago, "When we try to pick out anything by itself, we find it hitched to everything else in the universe."

Why, indeed, did we lose that perspective in the first place? Not many make science the scapegoat; there is, after all, a deeply entrenched pride in man's scientific competence. It is more the fashion to accuse the technological behemoth and to ignore the obvious--that technology per se can exert no value judgment, that both the prerogative and the responsibility for judgment lie in the minds and hands of those who create and use the tool.

Some two decades ago, soon after the advent of the nuclear age, we envisioned the birth of a new science--the science of peace. Our sense of urgency has not diminished; if we are to survive in a system we agree is closed, we had better create a scientific discipline of peace and do it quickly. Since the publication of *Towards a Science of Peace* in 1955,[7] the previously mentioned peace research centers have been established, along with several peace research journals; the demand for peace is made with increased resonance; the United Nations continues to function, at times precariously, yet with a basic measure of tenacity. We need, however, new impetus, renewed vitality.

We are prepared to midwive a new birth--to invest some life into a technology of peace and foster it on its way. This new concept represents not a break, but an extension of our previous position. We recognize the semantic difficulties of the phrase, "technology for peace," but rather than apologizing, insist that a technology can be sensitively constructed and validly directed to human need--not only to physical reinforcement but also to that part of human need which is social and conservative of collective life.

We commit ourselves to a hard look at the effectiveness of our current efforts to achieve peace and to devise needed improvements, beating out a gang plow of faith which attests to the possibility of improvement and the applicability of a technological approach.

Assessing the increased need for selectivity in choice,

Alvin Toffler says in *Future Shock:*

> Today we need far more sophisticated criteria for choosing among technologies. We need such policy criteria not only to stave off avoidable disasters, but to help us discover tomorrow's opportunities. Faced for the first time with technological overchoice, the society must now select its machines, processes, techniques and systems in groups and clusters, instead of one at a time. It must choose the way an individual chooses his life style. It must make super-decisions about its future.[8]

There is, without doubt, due reason for a widespread ambivalence when the value of technology is assessed. Automation has gone far to relieve drudgery, yet we have not been able to conserve adequately the value of personal creative involvement. Transportation and communication systems have abbreviated time and space, snugly binding the meridians and parallels of the earth into an uneasy and mistrustful intimacy. We have warmed and fed a portion of the world into a state of unprecedented comfort, polluting it, all the while, with the by-products of the process, and infinitely more tragic, leaving the remainder of the world in a state of desperate want. We have extended the quantity and enhanced the quality of individual life beyond our most imaginative dreams. At the same time, with our weaponeering, we have made possible the final annihilation of all life on our planet.

If we are afraid of technology, we must blame ourselves for that fear, for we have not had the courage and imagination to extend what we have discovered into the area of human and social relationships. In our plea for a science of peace, we suggested that the cure for the ill effects of science can be found in new and better directions; the same principle holds for technology.

The peace technologist will have to discipline himself to relevance and to applicability; he will deny himself the luxury of disorganization and search out the points of impact where knowledge may be applied against the political, psychological, and other institutional barriers standing firm between present reality and the vision of a warless world. He will have to identify existing gaps in that knowledge, envision the tools and techniques needed to fill the gaps, and modify current ineffectual techniques--all to the end of achieving a global unity.

We shall, in our discussion, first expand our rationale for urging a peace technology, examine thereafter its probable evolution and implementation, and finally—putting the technological horse behind the cart—address ourselves to the necessary antecedents to a technology of peace.

We hope the reader will not insist upon a too distant view. We know the forest is there--a forest that we hope will luxuriate beyond our most passionate hopes. But in the meantime, we are looking to the trees--to the quality of the stock and the condition of the soil and the prudent management of all the life forms in that forest--for the eventual health and vitality of the whole.

The reader will be asked to tolerate some repetitions. All our topics are firmly interrelated and likely to stay that way! We hope that a certain amount of repetition will serve to reinforce and emphasize, to the service of the reader.

We know how to wage war with appalling variety and increasing efficiency of method. We are, however, thoroughly fragmented and poorly equipped to secure peace. The imbalance can be corrected if we give careful attention to what we have not done; here, as elsewhere, our sins of omission are greater and more critical than our sins of commission. Peace-minded people know what they are against, but consistently have difficulty determining how to implement what they are for. If we profess to be committed to the total abolition of war, we had better be about the gritty task of getting the know-how for peace.

REFERENCES

[1]The *Peace Research Abstracts Journal* is published monthly and edited by Alan and Hanna Newcombe (Oakville, Ontario: Canadian Peace Research Institute).

[2]Anatol Rapoport's article, "Can Peace Research be Applied," is found in *The Journal of Conflict Resulution*, XIV, 2 (June 1970), pp. 277-286.

[3]Alvin Toffler is quoted from *Future Shock* (New York: Random House, 1970), p. 383.

[4]From "Ecology Manifesto: The Power to Create," *Our Generation*, VII, 1 (March 1970), p. 99.

[5]From an article by Paul R. Erlich and John P. Holdren, "Technology for the Poor," *Saturday Review*, LIV, (July 3, 1971), pp. 46-47.

[6]The original reference for this is from a poem by Langston Hughes, "Harlem," *Montage of a Dream Deferred* (New York: Henry Holt and Company, 1951), p. 71.

[7]Theo. F. Lentz, *Towards a Science of Peace* (London: Halcyon Press (Barnet) Ltd., 1955).

[8]From Toffler, *Future Shock*, p. 383.

2. Characteristics and Foci

The probable characteristics of a technology for peace include experimentation, measurement, the channeling and conversion of energy, the testing of old and discovery of new tools and techniques, the improvement of institutions, and the development of new institutions and mechanisms pertinent to bringing about the cultural conditions which will favor global unity.

Peace research, as we have noted, has made progress. One should criticize the peace research movement with restraint, considering with what limited manpower it is staffed, yet without a technological orientation the research being done cannot extend to a level which insures applicability. We hope to amend this by an overview of a technology for peace which will not only identify the shortcomings in current methods, but also propose the sort of technological orientation which will allow relevant experimentation and realistic goalsetting.

What are the first steps? The differentiation of idealism simple, and idealism hitched to a strong sense of realism, is crucial. The foci must be wisely chosen and the methodology perfected. As all the parts fit together more smoothly in the new technology, peace researchers—able to apply their findings—will function with reinforced morale; reciprocally, technology will be able to utilize the rising tide of peace research.

Experimentation is necessary to test new techniques. Measurement is essential for greater accumulation of facts useful to planners; it is also the launch pad for highly pertinent theoretical research. We must test our old measurement tools and invent needed ones so that we will stop guessing about the quality of peace interest and about what furthers or stifles motivation for peace.

Technology has vigorously furthered the efforts of physical science with its impressive harnessing and conversion of various sources of energy. How did prehistoric man get on without windpower, waterpower, or steam--forces untapped for tens of centuries? Where are the unharnessed energies awaiting release for the service of peace technology? Abundant but inoperative energies need to be identified, harnessed and channeled. Such energy potential as manpower and moneypower is obvious and self-definitive. There are other more subtle sources. There is a reserve of moral and emotional parental affection in most humans, and the importance of peace to the welfare of our children and each generation in turn would seem to be obvious. How then can a significant portion of this affection of parent for child be harnessed and transferred to an effective love for mankind?

Knowledge that is lacking but indispensable must be envisioned first and then acquired. Research must be structured with the qualification of relevance constantly in mind. Even when relevant, new knowledge should be tested for applicability. If the test is passed, better techniques are required for application. Experimentation to prove the usefulness of new techniques--and some which remain unused because of skepticism--will necessitate the investment of manpower and moneypower in new dimensions and directions. It is a step-by-step, interlocking process, which will require far more patience than bravado!

So to the search for relevant areas of impact! Peace technologists must determine and focus upon those points within the cultural complex where coordinated search appears feasible--and essential if we are to avoid the extinction of our species.

The defects in our culture, including the deficiencies in our institutions, through which individuals or groups must work, must be pinpointed. There is a distinction to be made among the various blockades to peace--some are pertinent to immediate and some to future progression of research and to the application of research findings. That this distinction has not already been made is due in part to the lack of proper measuring tools needed for doing the research necessary for appropriate judgment. The shortage of working tools, therefore, needs immediate attention.

It will undoubtedly be frustrating to decide upon the appropriate focus for research or for the application of results already available. Those who take on the work must have courage and patience and anticipate an evolutionary pattern of development. As the technological process evolves, the various early steps will undergo change; as a result the focus may change. For example, a shift from negative to positive in general emphasis should effect a shift of attention from military establishments to constructive international institutions such as the United Nations.

An important part of formulating a technology for peace is the wise choice of working assumptions. The danger of wrong assumptions can be minimized if made explicit in all intercommunications among the various scholars and thinkers in the field. The list of topics which make up this treatise contains some positive assumptions of prime concern for the creative peace researcher. These assumptions are germinal and must be criticized as such for needed deletions, additions, and modifications, according to the best criteria in the scientific and technical process.

One crucial aspect in viewing assumptions critically is to decide which of the old ones are most worthy of our testing time and by what method. Laboratory testing for the most part is too easy and simple an approach, the results unimaginative and inapplicable.

It may be profitable to give partial ear to the voice of the cynic who contends that the scientific method is not applicable to the subtle processes in psychology, nor to the macroproblems of sociology, political science or international relations. We hold firmly to the view that such application is both possible and essential; we also are sure that it requires imagination and patience in good measure! Already we have seen a few breakthroughs in social science.

It appears there has been relatively too much empirical and too little theoretical research. Certainly it should be possible and profitable to make essential shifts from the small laboratory to the larger one without abolishing the essence of the scientific attitude in testing the pertinence of an assumption. Obviously the most difficulty will arise when we apply the principles of experimentation at the level of the whole nation or the globe. Nonetheless it is stimulating to speculate on the possible use of satellites in carrying out global experiments--for the collection of data--for feedback to the researchers--for the dispersal of acquired knowledge from research teams to the general public. All this for the more distant stages of peace technology--but realistic.

Until that time, some cautions. For the moment, lacking sophisticated methods for large-scale operation, the bite-sized principle is the better choice--one which may be followed through to applicable or at least visible result. The testing of assumptions should be selective; at the outset we must not insist on clear-cut validation for every assumption.

One goal of peace technology should be the establishment of new institutions as well as improvement of the old. The failure of social institutions to keep up with the changes brought about by physical science and technology points up the need for focusing on the continued development of a social science and technology. The list for institutional focus is almost overwhelming--political, educational, scientific, economic, communicational. Need for change in these areas is obvious if conditions on this planet are to be improved—conditions that bring continued poverty, ill health, hunger, pollution, gaps between rich and poor nations and war.

Several studies in the field of peace research have already been made for the purpose of delineating the world of the future--studies on political integration, disarmament, economics and peace. Essays and several books have been written describing possible theoretical research on the future.[1] The results indicate that we will need to use the greatest reaches of imaginative insight in the search for remedies to these subtle, elusive problems. Usually the futurists have been content with mere extrapolation--the extension of current lines--with not enough emphasis on what could or should be. The goal of humanistic technology is "to make the most desirable the most probable."

The concept of attitude is another significant point of focus for the peace technologist. Much has been said about the blockades to peace. What are the attitudinal blockades whose removal will eliminate the economic and political barriers to a warless world? Very few studies have been directed towards answering this question. Peace actionists, who seem to assume that we know the way to peace, nevertheless move awkwardly, attacking the problem with imprecision. Peace technology will need to move towards quantitative validation of the assumption that everyone possesses the will for peace, asking by whom and to what degree of intensity it exists.

There is great challenge in technologically oriented attitude research and response to this challenge will have a high priority on the working agenda of peace technologists. They will be asked to identify and utilize attitudes already in existence. Or they might focus on the elimination of current defects in attitude research, thereby clarifying terms, agreeing on definitions, eliminating useless repetition of studies. Data analysis must be improved. Judgments of relevance must be made. If there has been too much concern with attitude towards war and not enough with attitude towards peace, the focus would be on ways to make an appropriate shift from the one to the other.

Researchers must nourish an enduring passion for facts relevant to peace. The technologically oriented will not be content with unqualified generalities, such as the belief that everyone wants peace. They will be concerned with the geographical and populational spread of this desire and its degree of intensity from person to person.

The frustration in peace efforts in the past has been

partly due to guesswork on the sources and resources for peace action. Reinforced by better data the technologist will concentrate on eliminating the frustration of the researcher and the peace actionist. A low level of morale has been served by the belief that what the individual does is of no account, or that no one else wants peace enough to invest the necessary time or money in the problem--or most importantly, that the problem of war abolition cannot be solved. If the arousal of public opinion in the masses has been too difficult in the past, then technologists need to focus on the identification of the submass groupings that could be effectively aroused.

Even the beginning student of peace problems, or the researcher, or technologist, should develop his own points of focus. With such guidelines he will keep his bearings in preliminary research and will choose increasingly valid pivotal points of value for starting himself in an imaginative technological or scientific direction.

REFERENCES

[1]Some suggested readings in the area of futures research:

(1) Robert Jungk and Johan Galtung, eds., *Mankind 2000* (Oslo Universitetsforlaget: London, Allen & Unwin, 1969). This publication is Future Research Monograph No. 1 from the International Peace Research Institute, Oslo, deriving from the presentations at the 1st International Future Research Conference in Oslo in 1967.

(2) *The Futurist,* published bi-monthly by the World Future Society: An Association for the Study of Alternative Futures, with headquarters in Washington, D.C.

(3) Paul Smoker, *Simulation for Social Anticipation and Creation* (Northwestern University, Evanston, Illinois, 1969).

3. Peace Education

The goal of peace education technology will be to create an educational setting which will stimulate specific progress toward a world without war.

Educators often claim that peace is the inherent goal in all learning endeavor. It would hopefully be true that a well-educated person is led by broader insight to a deeper understanding of humanity's needs. It may be that this hope is too subtle an approach to a problem which, in view of the risk, requires a more dynamic strategy. The goal of peace, as it stands in education today, is too often simply an inference, an unimplemented hope, a vague assumption.

There are other and more commonplace assumptions in *general* education that we must guard against, lest they be carried over into our thinking about *peace* education. We tend to give unwarranted priority to literacy and generalized knowledge, perhaps forgetting that education is orientation to living, and as such, must provide some specific guidelines if that living is to be carried off successfully!

Knowledge contains enormous potential for power and we can err dangerously by transmitting it into irresponsible hands. William James commented, nearly a century ago, that he knew of no instance of organized rascality in Eastern Massachusetts that was not initiated and promoted by a Harvard graduate. Assuming that Harvard has not retained the exclusive die for producing intelligent rascals, we should guard against the university becoming a menace rather than a hope.

The thoughtful consideration of peace education will find that the objectives are divided into short and long-range goals. Failure to delineate the divisions may account for much of the confusion heretofore encountered in peace education discussion. We may be criticized for repeated emphasis on such preliminaries as the need for increased task force and expanded financial bases, or even for stressing the need for the immediate training of more peace researchers. But these are short-range needs and only one of the approaches. Longer-ranged objectives will focus on the modification of educational techniques so that world-mindedness can be more effectively developed. Even more essential among the long-range thrusts is the development of a generalized and deeply rooted sense of social responsibility, which will be impregnable to irresponsible argument.

A statesmanlike approach to the technology of peace, including peace education, begins with the goal of abolishing international warfare. The starting point is the formulation of a peace philosophy and theoretical anticipation using suspended judgment; the follow-through will be careful research to see more accurately what is required. In this way objectives will become clearly defined.

Why do we speak here of a "philosophy" of peace? Ideally, philosophy is a forerunner of science and when properly directed, will provide some guidance for science;

that is, the humanistic and practical-minded philosopher can tell the scientist what to look for--or suggest approach procedures and appropriate guidelines. A well-defined peace philosophy is needed in both peace science and peace technology; conversely, philosophy itself can profit from a measure of technological orientation. This implies a division of labor among the technologists, with one sector concerned with the theoretical aspects of research.

In the task of defining objectives, a look at the history of science and technology might help the philosophers of peace to identify prior miscarriage and to circumvent the duplication of error. Then, as objectives become clarified, the directions for recruitment and curricula development will take form.

If effective recruitment for an increased task force is to begin, those persons responding must be offered specialized vocational guidance and clearly defined peace study curricula. The education of the person who chooses peace as a career will be a crucial factor and will demand much realistic thought regarding procedure, content and financing.

Recruitment is not simply the haphazard find of an occasional person who is willing to be involved in peace as a vocation. Effective recruitment is always predicated on good organization, and we have very specifically failed to provide the kind of organizational strength that changes the impotence of the isolated man to the potency of collective man.

There is general agreement that our survival depends heavily on the medical profession and we have medical schools to train physicians. We accept automatically the necessity of administration in the military and have service academies and officers' training schools. At what point will we feel with equal urgency the need for peace leadership and therefore establish academies for peace?

The exigencies of economics will compound the problem for the educator who, however much he desires educational reorientation, will nevertheless have to confront the task of financing his program. Getting monetary support is a competitive business. Social and philanthropic organizations--educational, religious, economic, political--have not allocated funds to programs of peace studies and research. It is very likely that with adequate financing enough individuals, already highly motivated, could soon acquire the skill needed for peace technological action, including peace research.

What is the present status of peace in the educational panorama? An increasing number of colleges and universities are offering isolated courses. A few schools have departments of peace studies. This is good, but limited, progress. Of the 10,000 students covered in these curricula, perhaps one percent may be taking peace studies seriously enough to decide for peace as a full-time career. Peace studies must be not only the antecedent but also a continuing dynamic of an effective peace vocation. Much of the approach presently used is interdisciplinary. We need the skill to determine whether this approach is sufficiently productive of peace attitudes, or whether the problem is better served by the development of an entirely new discipline.

Probably both methods will be effectively used. With research and restructuring, established courses can be peace oriented and new ones initiated. Professor Dwight Arnold, at Kent State University in Ohio, has done excellent work in this area. Good, existing interdisciplinary programs should be retained with an eye to constant reappraisal; in other circumstances, the independent department may carry the most thrust. Quality, not variety, will be the more absorbing problem.

From a longer proposal for a peace studies curriculum, some few areas of interest are: the history of international conflict and cooperation, the psychology of attitude toward the United Nations and world government, the economics of peace, mass communications and propaganda in relation to war and peace, the role of education in perpetuating war and promoting peace, relation of cultural aggression and resistance to peace, the principle of reciprocity in inter-group and inter-nation interaction, the principle and role of neutrality in conflict adjudication, the problem of language--universal and auxiliary, the relationship of such factors as population, race relations, and ethnic animosities to world peace.

Some of the curricular objectives, interrogatively stated, might be: what are the value conflicts that hamper peace, what economic interests condition the progress of peaceful cooperation, under what conditions has international cooperation developed and why have other peace movements failed, what are the potentials in religion for human unity and disunity, how democratize the science of peace, yet avoid research leading to an Orwellian 1984.

There are some related considerations which, although not inherent in the educational process itself, are crucial to its effectiveness. Whom should we educate first--do we begin with the diplomats and the politically powerful, the masses, the young, or the educators themselves? What stress must be put upon full-time commitment? As peace education develops, it must become less haphazard and be, instead, an everyday affair--everyday, not for everyone, but for an increasing number. We have observed the efficiency of division of labor. An expertise in peace technology will require not some effort by all persons, but all the effort of some persons.

How can we raise to full efficiency the techniques of inter-communication? When we have identified the persons strongly attracted to the concept of peace technology, they should be put into communication immediately. Just as engineering progress has demanded the interaction of the physicist, chemist, and mathematician, so technological progress in peace will require, and profit by, whatever is appropriate on the shelves of the social scientists.

There is an abstract knowledge about peace and war. We must discriminate between that knowledge and the kind that will favorably affect behavior and identify peace technology as the road to global peace. Peace education will lead to an identification with humanity and a commitment to the whole of humanity. But it takes a concentrated, preplanned and directed effort.

Scholars, researchers and educators with high peace interest and high morals must work together on developing a new technology: devising ways of implementing or applying research already done, outlining new research projects, inventing new tools for educational experimentation, and proposing other innovations for educational organizations. Ideas and questions will flow from the work of such groups into all areas related to or interwoven with education in its broadest sense. Science and its helpmate, technology, can then make discoveries that will be usable in generating a change toward a social action for survival.

As a technology for peace develops, it may become a prime educational force in its own right. This might eventually make much present day thinking on peace education irrelevant and raise serious questions not yet clearly visible. This is the challenge to our pioneering courage. The bear did not know, when he started to climb, what he would see when he reached the top. Not only was there the other side of the mountain, but beyond that, other mountains yet to be climbed.

4. Peace Politics

A genuine technology of peace politics requires that politics be conceived as something much more than governmental. It will include man's every organizational enterprise--government, education, economics, communications, judicial and religious systems. By exploratory, experimental and empirical research, peace technologists will identify necessary changes in our institutions, so that the will of mankind for greater harmony among men and nations can be more effectively expressed.

Here, as in every other facet of a proposed technology of peace, we assume a basic desire for peace in all individuals. Disillusionment and frustration have so dwarfed this desire that the cynical insistence that people don't actually want peace is very nearly validated. This trend is due a reversal. Tangible opportunity must be provided for gradually increasing numbers of persons to express the will for peace overtly through appropriate institutions. There may well be widespread mistrust of institutions among the people; indeed, it has been observed that our biggest crimes are committed by institutions rather than by individuals. But Kenneth Boulding, the eminent economist, suggests that the greatest error of omission on the part of the individual is his failure to help improve existing institutions or to bring about more relevant ones.

The individual can best express his desire for peace through collective action and collective action requires organization. Individuals must have organizations or institutions through which to cooperate toward a common goal. Institutional processes must be examined to determine where new structures have to be designed and which of the old ones may be effectively modified.

We would suggest some additions and improvements in several general groupings: politically, a strengthened United Nations and a National Peace Agency in every country; scientifically, a revamped UNESCO, primarily devoted to war elimination and new private peace research groups; in economics, research-granting peace foundations and a stronger World Bank to relieve economically engendered political tensions. We need educational facilities with a new orientation--world universities, university peace departments, and colleges for peace engineers. With improved communicational institutions we can discover, develop, and operate techniques and mechanisms for improved mutual understanding among nations and nationals. Other structures–the judiciary, the religious--could be modified to bring about a creative, cooperative society.

B.F. Skinner contends that man's escape to freedom is through the control of his environment.[1] It is well to add that the control of man's environment is achieved by the control of his institutions. Karl Marx said that history is the actions of men in pursuit of their ends. Faulty institutions and individual attitudes make it inevitable that the pursuit of those ends will result in unfortunate social outcomes.

The obsolescence in many of the governments throughout the world is a matter for serious consideration. The misuse of parliamentary lobbying, the reliance on military power for national security, the entrenchment of defective electoral systems, and the extremities of partisan politics are all problems that should challenge the peace technologist.

Commercial imperialism and its attendant pressures on many groups--scientific, educational, communicational, governmental--may constitute one of the knottiest problems for the peace technologist to resolve. The phenomena of world-wide commercial corporations requires and deserves more thoughtful, open-minded and constructive consideration than thus far received. Shall these burgeoning national and international organizations be reconstructed, and if so, what modifications will succeed? If substitution is the only viable answer, what will be the nature of the replacement? The best solution may eventually prove to be control by international organization--world government--which in turn would present a renewed challenge to peace technologists. Both theorists and other researchers will need to envision techniques for implementing some form of world government through which the ultimate goal of human survival can be realized.

The United Nations and its many subdivisions should be considered both for critical improvement and creative expansion. A revamped UNESCO could be our key instrument for a specific approach to peace-a clearing house for present and future peace research on an international scale.

In recent years in the United States a growing concern about priorities has been evident. The Senate

Foreign Relations Committee has conducted extensive hearings about the Vietnamese war. New alignments have been shaped by the formation of such groups as the Student Non-violent Coordinating Committee and Students for a Democratic Society. The number of peace courses offered in colleges and universities has risen, and not least, there is the threat of revolutionaries. Even revolutions often result in the substitution of one defective government for another. All these groups and mankind in general will remain essentially helpless unless new institutions are made functional, wherein systems of social goals can be defined and implemented.

We need world universities, departments of peace studies and peace research. We need colleges of peace engineering and law schools primarily devoted to world law. We need political science departments that focus sharply on the goal of political global unity--curricula in which the principles of power politics are supplanted by the principles of cooperative politics.

We need new communicational structures to discover, develop, and operate techniques and mechanisms for improved mutual understanding--among individuals and groups--within nations, across national borders, and among nations. This will necessitate a critical and constructive attitude toward mass media in its present form. As more people are involved in decision making, mechanisms for speedier feedback will be required. We suggest the use of the term "intercommunication" rather than the more common "communication." How else can we increase the awareness and cooperation of widely scattered persons?

We need a worldwide economic system, created for valid moral reasons, to erase the disparity between the poor and the wealthy. A strengthened World Bank could relieve economically engendered political tensions. War bonds were successful by the dual appeal of psychological and economic profit. Peace bonds could function in precisely the same way. Research has indicated that a peacetime economy can support a super-industrial system as well as can a wartime economy. Kenneth Boulding and Emile Benoit have edited an extensive investigation into the effects of disarmament on the economy.[2] The latter, in a paper on the subject, suggests that if we can re-educate the people to new responses to government spending and indebtedness, the level of demand can be maintained and recession avoided. If demand can be maintained, the structure and redistribution problems can be resolved. In short, *the real obstacle to disarmament is not economic, but political.*[3] We had been convinced otherwise in the past; technology should tackle this and other blockades in our thinking.

A better range of measurement tools would result in less pretense and more of the kind of factuality needed to guide mankind through the political mazes of today's world. It seems that men's major efforts are made to make an unworkable system work; precise tools would reduce the guesswork about the efficiency of current institutions and individual efforts.

Relationships existing between the concepts of science, technology, and engineering must be carefully searched. No clear-cut boundary can now be drawn, except for purposes of extreme simplification.

Eventually experts schooled and skilled in the theory and practice of political and organizational thinking will be able to make a better exposition of the technology of peace politics. But we can say in summary that the purpose of peace technology is to escalate the peace movement in its political and institutional aspects, replacing the emotional war system with a reinforced, rational system of peace.

REFERENCES

[1]This thesis is developed in B.F. Skinner's *Beyond Freedom and Dignity* (New York: Alfred A. Knopf, 1971).

[2]Reference is to research by Emile Benoit and Kenneth Boulding, eds., *Disarmament and the Economy* (New York: Harper and Row, 1963).

[3]From a paper by Emile Benoit, "The Economic Impacts of Disarmament," *Behavioral Science and Human Survival,* edited by Milton Schwebel (Palo Alto, California: Science and Behavior Books, Inc., 1965), pp. 111-124.

5. Attitude and Motivation

Central to the goal of peace technology is the recruitment of a task force whose strength lies not only in number, but in the possession of clearly defined attitudes and equally clear concepts of how these attitudes may be translated into action. For the individual this means the translation of genuinely conceived peace concern into specific and feasible effort toward the abolition of war. We should define the method as precisly as we are able, given the human tendency to imprecision.

So the development of a technology for peace requires the consideration of how and when attitude results in overt behavior and how that behavior is made evident.

Human activity does not spring full-blown from human attitude. Attitude is a behavior tendency with varying degrees of intensity until some activating factor is added. Here the catalyst is opportunity; it is the stimulus which converts tendency into motivation and, in turn, into overt behavior.

This is the process, then, which engenders mature behavior; and by examining its several functions, we can discover the areas to which attitude research must address itself if it is to be fully effective in the area of peace technology. Research must determine what attitudes do exist, to what extent and by whom they are held, and their possible correlates and causations. Further investigation should identify procedures for effectively influencing and encouraging the development of these attitudes and for determining the conditions under which they will be translated into behavior.

Attitudes toward peace must be discovered, where latent. We shall need not only to be able to identify this potential, but to apply some tools of measurement to determine the degree of intensity. It will be necessary to have a variety of methods at our disposal. The interview process allows the surfacing of unknowns: on what terms do persons want the elimination of war; to what degree are peace attitudes latent in the intellectual and moral elite; is there a quiescent willingness to provide financial sinews for burgeoning movement. If we find there are persons predisposed to a technology for peace, how do they envision appropriate direction for this technology; is there a tolerance of diversity and less emphasis on individual difference, and is there a viewpoint extending beyond nationalism and recognizing the possible requirement of a new international system.

Interviews spanning a score of years have given me the strong impression that many genuinely concerned persons appear apathetic toward the threat of annihilation by war, but that in fact, this is less apathy than the inability to identify viable alternatives. "What can I do?", is a plea for direction, rather than a protest against involvement, and this is an insight gained more easily in the personal exchange of an interview.

The overview thus acquired will function as a point of departure into more specific processes. There is a spectrum of tests and testing techniques. Perhaps the simple questionnaire would have its usefulness, or the opinion poll might be further adapted. Social scientists have discovered the applicability of sampling techniques to huge populations; a well-chosen sample for a simple opinion poll of 1000 persons can tell us with great accuracy what we would get if the same questions were put to one or two hundred million people. A sample of 25,000 persons would probably give us an accurate measure of the opinion of the total world population. A sample of 50 to 100 words will give us a good measure of an individual's vocabulary, facilitating the measurement of the vast psychological storehouse in a single mind. More empirical work toward greater factuality in measuring peace technology attitude will be required, and we have made a promising start on such a tool in our research in St. Louis, which we call the Peace Technology Attitude Opinionaire.

The opinion poll may prove effective, yet may itself wait for further extension toward the depth interview in which we will have the manpower to ask fewer people more questions focused on a central problem. A strong peace technology will not discount the immense possibilities of this breakthrough in social technology.

What factors are there which produce attitudes toward peace, rather than toward aggression, militarism, or factionalism; what situations nourish and stimulate them? One attitude research finding in a report by William Eckhardt points up, at least hypothetically, the advantages of growing up in a permissive situation in home and school,[1] but the researchers have not gotten to the point of knowing how to define permissiveness in such a way that specific modification can be applied to specific circumstances. We can assume that leaders in peace research centers will become disenchanted with research findings that are not applicable; some are, in fact, already demanding applicability. A zest for facts, objectivity, discontent with established criteria, semantic curiosity, a dynamic concept of attitude development--all these are traits which must be encouraged for our purpose. What milieu serves that purpose best?

Influences such as morale, habit, inertia, apathy and personal gratification are all bits of phenomena related to peace attitudes; the individual must be led to convert unenlightened self-interest into enlightened self-interest. The latter assumes an involvement in the whole of which he is a part; it is a shift of interest which will require interpersonal skills as never before. The realization will come that attitudes change less through verbal persuasion or other coercive methods than through the pursuit of successful and gratifying cooperative activity.

As the technology of cooperation evolves, methods will develop by which the potential cooperator may acquire a facility for self-criticism. With new clarity he will be able to appreciate his attitudinal handicaps, and at the same time, to identify commonly held concerns, a vital factor affecting the motivational shift to concern for humanity. The shift itself will require increased use and refinement of techniques for intercommunication. How many persons are in touch with one other person dynamically concerned with the scientific and technological approach to peace? The role of morale should not be minimized; the discovery of like-minded persons and the growth of intercommunication among them should boost morale appreciably.

A crucial interaction should spin off. There is at present a very real distrust of technology, surely due in part to technological miscarriage, in turn due to inappropriate motivation. When the partisan motive controls technology, we get polluted air and rivers and oceans, selfish aggression and thwarted relationships. But if tools for better communication produce a motivational shift toward corporate concern for the welfare of the world public, we will have to acknowledge that those tools have come out of a cogent social technology, reinforced by success, and further technological improvement will be stimulated.

Peace research should start with the theoretical, and produce hypotheses followed by experimentation and further observation. Technological attitude should similarly be aware that both our culture and our technology are lopsided, but that the imbalance is correctable; it must envision corrective measures. It will have not only to apply old scientific knowledge currently at hand, but also to develop new knowledge which will weaken or destroy political and economic and psychological barriers to the realization of a warless world. It must present credible opportunity for peace relevant behavior, exercising merit judgment as well. The vast amount of activity in the United States directed toward the ending of the Vietnamese conflict was much closer to our need than a recommendation to blow down the Pentagon. The implementation of a demand for a Department of Peace within the national government would seem to be as pertinent at the moment as continual diatribe at the military-industrial complex.

The United Nations faces extended crisis. Is there a deep-lying antipathy to the concept of world government, or any other world authority, that renders its influence unavoidably tenuous, or is the crisis essentially financial? For each dollar paid by the United States taxpayer toward the misnamed Defense Department, only a fraction of one cent of another dollar supports the United

Nations. We must discover whether the priority reflected can be altered simply by education to the need, or if deeper attitudinal changes are required.

Peace impulse is closely linked to the growth of peace technology. One usually wants to do what is considered feasible. The urge for transoceanic and transcontinental travel grows with jet flight opportunity--a matter of both technology and finance. Space travel moves out of the pale of science fiction only as technology makes space transportation feasible.

We cannot avoid the bugbears of size and cost. There is an appalling discrepancy between the size of our task and the miniscule size of our task force, and peace can be achieved only by the collective action of sufficient numbers with viable peace attitudes. These attitudes, at the outset, may probably be found in limited numbers rather than in the masses.

However, one tenth of one percent of one billion equals a million, and a million people, adequately intercommunicative, will carry an effective clout! We said in our first book:

> Peace is indivisible. It is international peace that is discussed in this manuscript, i.e., world peace. When it comes it must come to all. It must exist universally or not at all. This is not true of its causation--at least not the scientific beginnings of its causation. Honest decision to contribute can be made anywhere, anytime, by anyone with concern and social imagination. Democratic disarmament will require wide unanimity. Not so with disarmament research. The challenge of the universal ignorance which tyrannizes over us all can make its first appeal to a sensitive and alert few. These few can save the

many if they be confident, and if they intercommunicate.[2]

We are a world population acquainted intimately with economic exigencies; we devote the greater part of our time and energies to vested interests, in the broadest sense of the term. Yet we fail to recognize the economic value of peace and continue to spend billions of dollars for armament and comparative pennies for disarmament. There is an hiatus in attitude development here that leaves the individual unprepared for sound peace stewardship, and carries through to the level of national and international priorities.

So we must deal with a crisis of faith. We do not have to be reminded that history is marked deeply with the phenomenon of one war after another and that eventually the peace technologists, as well as other serious students of the problem of military conflict, can afford to settle for no less than the total elimination of the war system. But we must believe, in fact, that it is possible to achieve permanent peace. When we are convinced that attitudes demanding peace may be identified and implemented, we will provide credible opportunity for effective action. When we have the tools and techniques to disseminate knowledge and then to apply it, we will liberate mental and financial energies. The wheels of peace technology will roll when we are willing to pay our way out of the Stone Age of military conflict, and when the emotional energy of hate for war can be transformed into kinetic energy for peace.

REFERENCES

[1]This is examined by William Eckhardt and Alan G. Newcombe in "Militarism, Personality, and Other Social Attitudes," *The Journal of Conflict Resolution,* XIII, 2 (June 1969), pp. 210-219.

[2]From Lentz, *Towards a Science of Peace,* pp. 192-3.

6. Discovery and Invention

One is, by this point, quite aware of the continual overlap of topics thus far discussed. Each facet of the problem of peace is closely bound to all others. This is extensively true in current peace research and will be even more so in the initial development of a technology for peace. As such technology does develop, however, we should be able to divide the field and conquer the problem of overlap. Meanwhile, we shall continue to repeat the need for new and better tools, techniques, and methodology in each chapter, since they are basic to the better social technology promoted in this book. It is hoped that through constant inventory of the limited choice of tools in the peace technology kit, additions to the kit will be assigned highest priority.

No technology develops without experiment; experiment requires measurement, and tools to make that measurement. Tools, in turn, have to be invented. In the physical sciences the task is exacting enough. But when the object of experimentation is human attitude and motivation, the problem compounds, and we are required to have enormous faith in and reliance upon the power of human imagination. The pioneer technologist must believe that the human imagination can operate skillfully and creatively in the area of social relationships as well as the physical. No less faith is required of the economic participant whose contribution of risk capital furnishes the services of the peace technologist, and in the broader sense, determines the limitations of the task force.

As the aggregate of available mental energy increases, more and more human genius will be diverted from lesser goals to that of peace. A second shift must also occur--from the negative goal of winning wars to the positive goal of human survival. The urgency of the altered thrust rests on the assumption that unless mankind gets rid of the institution of war this institution will put an end to mankind.

Creative pioneers can stimulate their own morale by studying some discoveries in the social field, even though relatively few in number and only partially developed. We have mentioned previously the refinements which have occurred in sampling techniques; we can scarcely pass over the impact of the computer on both the quality and the acceleration of our current social development. It is easy to think in terms of the computer as a product of the physical sciences, but in fact, it was originally an outgrowth of social science, a super-sophisticated extension of language and mathematics. It has, as Toffler puts it, become "a major force behind the latest acceleration in knowledge-acquisition"[1]--a pipe for the fueling of data--for "if technology....is to be regarded as a great engine, a mighty accelerator, then knowledge must be regarded as its fuel."[2]

Robert Boguslaw further develops the perspective that allows us to properly place the computer in an encompassing social structure.

> Computers are not found in nature. They have to be built. And they must take their places within a framework of existing social systems....(which) redefines existing system arrangements in significant ways....in information organization....formalized work procedures....and people.[3]

The role of focus becomes dramatic when we are aware of the brief, yet voluminous, history of discovery and invention in areas where the mind of man has been directed to a particular goal. Lewis Mumford reviews this facet of history in *Technics and Civilization,* and includes an awe-inspiring list of some 500 breakthrough inventions.[4] For those who are cynical about the value of our civilization, it is well to point out that dilemnas in general, and our particularized ones, are largely due to the imbalances in the channeling of creative and inventive genius. But balance requires patience. We learn to creep before we run. The spade preceded the bulldozer. The sailing vessel came long before the steamship. Steam itself was harnessed by Watt and his contemporaries without waiting for the invention of the dynamo and the motor.

Let us conjure up a dialogue between an electric dynamo and an electric motor which are in reality very significant teammates in electrical service--the production, distribution and use of electrical energy. In electricity, dynamo is to motor as peace technological companions are to X and Y. It must be the job of somebody to explore, first theoretically and then experimentally, for the X and Y. Both dynamo and motor are converters. The harnessing of steam energy precedes the dynamo and the transmission line precedes the motor. Of course, the motor is no good without a dynamo and though they are opposites, they are highly complementary.

THE DIALOGUE

Motor:

You make it and I spend it!

Dynamo:

What do you want to spend it for? Can you justify your needs?

Motor:

Sure, hundreds of kinds of tools are waiting to be activated by your energy product, through me—the motor, the motivator.

Dynamo:

Oh, are you a motivator?

Motor:

Yes, I motivate, or animate—anyway, put life into the inanimate.

Dynamo:

Are you the end of your dynasty?

Motor:

Not a very tactful question. But no, my offspring will motivate the apathetic human animal.

Dynamo:

Ah, you've got it easier than I! The animal is animate to

start with, but in the electrical field, you and the tools which you animate have no life to start with. I mean, the invention of a social or sociological motor should be easy.

Motor:

We'll see about that when we get dedicated, creative men on the job. But first *you* must produce an offspring that will harness psycho-social energy and feed it into the sociological motor. We'll bring up our progeny together.

Dynamo:

Too bad I can't go it alone, but that too will require men on the job, men with inventive imagination--that is, our reproduction will continue to require the intellectual energy of human minds.

Motor:

Correct! How might such men proceed?

Dynamo:

We'll have to figure that out after you motivate them to get on the job—sorry to keep coming back to that. Perhaps they can get some inspiration or clues from the early history of electrical technology. Maybe social energy already exists in greater unharnessed quantity than we realized.

Motor:

What do you have in mind for a clue? Perhaps you are thinking about steam and waterfall energy which existed long before it was harnessed and converted into electricity.

Dynamo:

Yes, and long before electrical energy was generated by human design, men were watching terrifying displays of unharnessed power as lightning severed the sky. And they were awestruck and more, for generations upon generations—first because it frequently reached the earth, destroying trees, structures and even human life, more especially because it was treated as supernatural or mystical phenomena and not as a riddle to be solved by human intellect. But Ben Franklin was marked by curiosity and *he* flew his kite toward the clouds.

Motor:

Good. But if we're going to fly kites, we at least ought to be systematic about it—do some careful and extensive observation. In the field of social and human relations, people are launching emotions into every passing breeze, but they don't know how to handle legitimate currents.

Dynamo:

That's nicely put. You're more fanciful than I supposed, but can you be more specific? What are some of the social energies that need to be harnessed for human good?

Motor:

Well, love and hate, hope and despair, trust and fear, love of humanity versus love of nation, pride in being human versus national pride, the spirit of cooperation versus the spirit of competition, the hope for humane use of nuclear fission versus the fear of total destruction by nuclear war.

Dynamo:

It might stimulate some thought to recall that volcanic eruption resulted from uncontrolled energy in the earth's crust long before James Watt and his contemporaries harnessed the energy of steam. Even now physical technology has not tapped much of the total force of subterranean energy. Vesuvius could be a visual symbol of man's incompleted work.

Motor:

So, like volcanic energies, psycho-political potential calls for more sensitive instruments to locate the eruptive forces so they can be harnessed for beneficent work.

Dynamo:

In our field, we have a wonderful triumvirate in the steam engine, the dynamo and the motor.

Motor:

Explain that. I was still thinking about Vesuvius.

Dynamo:

Steam supplies energy to the dynamo, which converts and conveys it to the motor by way of the transmission line.

Motor:

Who were the giants of discovery and invention in our lineage?

Dynamo:

Many giants. People catalogue them from time to time,

but for an instance—once upon a time, a Mr. Michael Faraday rotated a wire coil in a magnetic field and presto, electrical current resulted.

Motor:

Who discovered magnetism?

Dynamo:

That's another long and involved story.

Motor:

Why involved?

Dynamo:

Well, some of the beginnings preceded Faraday by 2000 years. Then the magnetic needle enabled a mariner to locate the direction of north on a very cloudy day or night. This magnetic indicator is credited as a great aid to Columbus in his quest for a new route to India, when he accidentally discovered a new continent.

Motor:

Can such accidents be prevented? No, I mean—*predicted!*

Dynamo:

No, however Columbus would not have made this particular discovery if he had devoted his time to winning a war for Italy or Spain.

Motor:

His motivations must have been complicated.

Dynamo:

That's so. And in the matter of invention and discovery, historians tend to simply record the fact—they don't get into the motivational and psychological conditions directly antecedent to the great discoveries.

THEY PAUSE TO INGEST SOME FRESH ENERGY

The line between invention and discovery may not be sharply drawn, but the process of locating it may challenge some preconceptions. One challenge is to the assumption that science *must* precede technology. Another concerns the limited definition of technology as the application of scientific knowledge. Dynamic technologists concern themselves not only with application of available knowledge but identification of knowledge gaps.

What are the conditions needed for breakthrough in discovery and invention? There must be an adequate number of men on the job. They will utilize time otherwise used for leisure or for remunerative employment. Buying their time requires financial resources. The increase of manpower will make possible a division of labor. Greater division and specialization of labor make for sharper focus. Enormous quantities of moneypower now flow in other and less important directions; harnessing a reasonable amount for peace technology will ensure the availability of imaginative energy.

Creativity is indispensable to discovery and invention. In the past technology has resulted from the freedom and dedication of innovative persons. So with the future of peace and survival. Latent dedication must be enlivened. The talents of persons whose skills are now directed to other and less socially important ends must be secured. The start will have to be made with the few that are most ready.

What other energies can we expect to harness? What about moral energy? Much rhetoric has been directed to the moral forces in human relations, including the impact on political decision making. Research suggests that a vast majority of persons are at a low level of moral development, though more studies are needed to support or refine this finding. It may be, in fact, that moral lassitude is due to lack of appropriate opportunity.

It is possible, therefore, that in developing or testing mechanisms or techniques for putting moral energies to work, an increase in such energies may result. There are oceans of parental love which are seldom used in any direct sense to bring about a warless world in which children might live without threat, though one would assume that goal to be of first priority. If we could measure attitudes toward world-mindedness, we might discover the clue to effective transition from nationalism to globalism. Anger has been linked to steam, but much of it is vented in rhetoric and personal accusation. If this anger could be turned against institutional defects, resulting action might be more positive in impact. Idealistic and restless youth might turn their attention from acts of destruction to a sensitive structuring of peace.

REFERENCES

[1]From Toffler, *Future Shock,* p. 31.

[2]Also from Toffler, p. 29-30.

[3]The role of the computer is discussed by Robert Boguslaw in *The New Utopians* (Englewood Cliffs, New Jersey: Prentice-Hall, Inc., 1965), p. 182.

[4]For this list, and an extensive bibliography on the subject, see Lewis Mumford, *Technics and Civilization* (New York: Harcourt, Brace and Company, 1934), pp. 438-474.

7. Manpower

To abolish war and attain permanent peace, we must increase the size of the peace task force to a level appropriate to the demands of the task. It is very difficult to foresee what this means in actual numbers. Some persons believe it must be as large as the force devoted to the waging of war, but one hopes that this estimate is overdrawn!

Whatever the requirement in terms of number, this treatise insists that the manpower problem deserves very high--perhaps the highest--priority and that we approach the problem with determination. The low manpower situation is not given much attention by the small coterie of peace researchers now operating in various centers, and it requires courage to emphasize this facet of the problem. To those whose focus has been narrowed to specific issues such as the Middle East conflict, the cessation of the war in Southeast Asia, the strife in Ireland, or the Pakistan-India conflict in the Asian subcontinent, it may be shocking that we insist on the primacy of the issue of manpower.

There is no question but that a large labor force accelerates the thrust in physical technologies. The total labor force for producing electrical energy throughout the world may well exceed a million workers. The number of stewardesses in the world airlines is about 50,000 and the total labor force in aviation is estimated at more than a million. On a worldwide basis there are more than 50 million persons employed by the military establishments.[1] The list could be enlarged, specifying the commercial and governmental enterprises that are sustained by huge labor forces and which have given the world a powerful technology which threatens to destroy it.

Compare the above statistics with the world's peace force, a pitiful number estimated at far less than the sum of airline hostesses. The *International Repertory of Institutions Specializing in Research on Peace and Disarmament* by Johan Galtung and Marie Rugi, published by UNESCO in 1967[2], lists two score centers throughout the world which are engaged in peace research. Some of these spread their efforts to include other causes and in at least some cases, the chief of the operation has another full-time job for the purpose of personal subsistence. One of the most dynamic centers, the Canadian Peace Research Institute, has only seven full-time workers.

If we examine the qualitative aspect of this limited coterie, we are struck by a contrast. The scientific scholarship is of high order; on the other hand, there is little evidence of technological orientation. After the first year of publishing his *Journal of Peace Research,*[3] Dr. Galtung commented that his goal had been two-fold--to find contributions of scholarly quality and of high peace relevance--and that he had succeeded well on the first count but poorly on the second.

One's reaction to a comparison of the relative size of task force in physical technology and present peace technology depends upon his perspective and courage.

The solution of the manpower puzzle is crucially tied to morale, and is one of the most daunting aspects of peace technology. Especially is this true when one measures the intellectual power now invested in enterprises of far less importance to, or even of specific negative effect on, the human situation.

As indicated previously, the question of the antecedent steps to the attainment of adequate manpower is of primary import. Sources must be identified; in the early stages, one will look for the most humanitarian elite. This surely will include deeply committed students of peace, but additional resources will be found among current ranks of the rebellious whose youthful discontent might potentially be aimed above or beyond the stage of destructive or non-constructive dissent. Their energies may be refocused to study the problem of peace intently.

There are social scientists who envision a revolutionary social technology as the answer to the future of mankind. In *The Moral Society a rational alternative to death,* John David Garcia says that "science and technology form a symbiotic pair"[4] and that the "only aspect of the total environment upon which the scientific revolution has not yet made a direct significant impact is the psychosocial."[5] Rejecting the descriptive ideological orientation of these psychosocial sciences, he emphasizes that "the point to be made at this time is that in all of human history only science and technology have been proven *directly* effective in expanding human awareness."[6]

Many physical scientists might be willing to re-direct their skills into research and technology for peace, even if some retraining were necessary. There is a potential in the ranks of housewives, mothers and the physically but not mentally retired. Dr. Elise Boulding, who launched the very significant Consortium on Peace Research, Education and Development, achieved a doctorate in sociology after her children were grown. Obviously one must be willing to return to school or work outside of academia in pursuing seriously and intensely the study of peace.

Even after human resources are identified, it will be essential to discover which of them are most receptive to the concept of a peace technology. An examination of the literature of the peace and peace research journals exposes the dearth of a peace technological attitude. The St. Louis Peace Research Laboratory has made a beginning in the development of an instrument for measuring this attitude--a Peace Technological Attitude Opinionaire.[7] It is anticipated that this can be a useful tool in finding or promoting the sources of new manpower for peace technology.

Suppose potential manpower is identified. How can it be enlisted? We intend this treatise to be a significant plea for help. It seems obvious that the peace analyst must be given a very important place in setting the goals for the society of man, but the analyst must be discovered and supported; this is a part of peace technology. New methods need to be devised, new tools invented, to find the potentials. In addition to the Opinionaire previously mentioned, the St. Louis Peace Research Laboratory has another partially developed tool, the Peace Career Expressionaire,[8] which, with broader circulation, may be effective in finding persons who are willing to work in this field.

As the numbers in peace technology grow, association for mutual aid will also increase. Intercommunication between members of any such group raises the morale of those engaged in the work. Cooperative thinking is useful in determining what is relevant research and could help in the development of a viable technological policy. There are at least two such associations already formed, the International Peace Research Association (IPRA), whose secretary-general is Professor Asbjorn Eide in Olso, Norway, and the Peace Research Society (International) under the directorship of Dr. Walter Isard, at the University of Pennsylvania in Philadelphia, U.S.A. There is also the Consortium on Peace Research, Education and Development, mentioned earlier in this chapter, now with headquarters at the sociology department of the University of Colorado, Boulder, Colorado, U.S.A.

Manpower increase will be considered as having two aspects--the qualitative and the quantitative--and there is valid reason to recognize the connection between the two. Since breakthrough in scientific research is partially due to chance, ten thousand searchers are more likely to find the right clue than ten. Among ten thousand dedicated searchers there is greater likelihood of the inclusion of persons of exceptional genius. Social science, as well as physical science, needs its Einstein, Szilard, Fermi and Oppenheimer. Large numbers of persons, appropriately organized and intercommunicating, are more likely to evolve cooperative thinking toward a technology of peace, or any suitable alternative that may be discovered.

There will be those who envision the necessary expansion in peace technology as coming concurrently with the development of social technology in general. But the problem of human survival comes first; other social goals may be more feasibly attained after peace is achieved. Should the human species become extinct, social goals will not make sense.

REFERENCES

[1]This figure is derived from information in *World Military Expenditures 1966-67* (United States Arms Control and Disarmament Agency, 1968), pp. 5-6.

[2]The *International Repertory of Institutions Specializing in Research on Peace and Disarmament,* edited by Johan Galtung and Marie Rugi, was published by UNESCO in 1967.

[3]Reference is to the *Journal of Peace Research,* edited by Johan Galtung at the International Peace Research Institute, Oslo, Norway, and published quarterly by the International Peace Research Association, Groningen, the Netherlands.

[4]The quotation is from John David Garcia, *The Moral Society a rational alternative to death* (New York: The Julian Press, Inc., 1971), p. 27.

[5]From Garcia, p. 28.

[6]Also from Garcia, p. 29.

[7]The "Peace Technological Attitude Opinionaire" was developed in the Peace Research Laboratory, see Bibliography.

[8]The "Peace Career Expressionaire" has also been developed at the Peace Research Laboratory, see Bibliography.

8. Moneypower

If the problem of manpower deserves top billing in the technology of peace, the problem of moneypower must rank second. With very rare exceptions, long-term and full-time peace workers will not be available without economic subsidy. Of all energies to be harnessed and channeled peaceward, financial power should rank high among our technological priorities.

The gross national product of one nation alone, the United States, was in 1970, 977 billion dollars[1] and the gross annual product of the world is somewhere around three trillion in dollars or dollar equivalent.[2] There is an enormous flow of money not only within but between nations for purposes which have little or no relationship to peace and disarmament.

It is painful to observe the small budgets on which the various peace societies must operate. Even those with the best support do not have budgets in excess of 3 or 4 hundred thousand dollars per annum. In 1969, one of the strongest of our peace educational and political units, World Federalists, U.S.A., had a proposed operation of $375,000. The World Association of World Federalists skimps along on a much smaller allotment and can support only five full-time persons in its world headquarters. These organizations, of course, represent rather exclusively the goal of world government as the basic solution to the problem of disarmament and peace.

In contrast, an estimated 200 Billion dollars are spent annually on world armaments.[3] As long ago as the 1930's there was much outcry about taking the profits out of war, and the outcry continues sporadically. Seldom, however, is anything said about putting those profits into peace, except in the most generalized terms.

In the United States, 12.5% of the earned dollar is consumed by Federal taxes, and the majority of the tax revenue goes to defense related causes. A breakdown of government spending seldom indicates the percentage of the taxpayer's dollar that goes to peace either on the national or international level. Perhaps this is true because the sum of money for peace is so miniscule compared to the enormous amount spent on the military budget. An increasing number of people in the world are making their wealth from war preparations. A much larger number--some 50 million--are comfortably salaried in the name of defense and security.[4]

The discrepancy between the appalling size of the task and the smallness of the individual is a psychological factor tending to inhibit the individual impulse to contribute. On one hand, we must certainly acknowledge the size of the challenge--on the other, we need an appreciable self-confidence and conviction of the potency of the individual. This is true not only in monetary considerations; the potential contribution of any person in any way carries with it the promise of effectiveness. Not all hope must be invested in those at the top of the political or economic power ladder; this would foster an unfortunate aura of dependence. The number of persons who can make peace a full-time career at present is small, however a few of our boldest have been willing to use some risk capital of their own. Many causes ask for support, but some few persons have given their personal priority to the cause of peace.

It is a truism that only the individual can act, but hardly deniable that his effectiveness is increased when he acts collectively. With this increase comes sharp changes in morale. The process of extending and collectivizing individual emotional and financial forces towards the attainment of peace and the survival of the species is crucial to the technological approach, humanistically conceived.

The motivation for financial contribution may have to be different--more philanthropic, in the best sense of the term--than financial participation in other technologies. Lewis Mumford, in *The Myth of the Machine,* speaks about motivation:

> In short capitalism exploited and universalized a powerful positive motivation that--for decent human reasons--had never been tapped in more primitive societies. For centuries, it is true, the capitalist continued to use the negative form, of penalization rather than reward, to ensure compliance by the worker, while self-righteously reserving rewards for himself, his managerial colleagues, and the investors.
>
> Money, as the nexus in all human relations and as the main motivation in all social effort, replaced the reciprocal obligations and duties of families, neighbors, citizens, friends. And as other moral and esthetic considerations diminished, the dynamics of

money power increased. Money was the only form of power which, through its very abstraction from all other realities, knew no limits—though finally this indifference to concrete realities would meet its nemesis in the progressive inflations of an 'expanding economy.'[5]

In all probability, we will eventually have to turn to taxation as a source of support--either within nations or by some international authority. If done internationally by an organization such as the United Nations, much procedural strengthening will have to occur first. The wealth of the oceans belongs not to private individuals, but to mankind. Why not tax it through a world authority for the benefit of all? If a national government is the taxing body, the administration of the monies is hampered by the criterion of national interest narrowly conceived. First in Sweden, later in a few other instances in Europe, international peace institutes functioning independently of their governments have been given small, yet significant, appropriations for peace research.

Preceding the growth of this kind of appropriation, however, we shall undoubtedly have to count on the philanthropic impulse of a few individuals with an appreciable degree of loyalty to mankind as a whole. We have maintained that "the acid test of peace interest is personal economic participation." Many say they can give some time but have no money--no money, apparently, for which there is no other use!

The hope for help from the large foundations has been singularly frustrated; these billions are not available for peace research. In 1961, the total capital of our foundations in the United States was reported to be about $17.6 billion. By 1971, the sum had grown to

$25.2 billion.[6] Throughout the world and especially in the United States, there are scores of philanthropic foundations supporting academic and scholarly projects, and research for the cure and prevention of diseases. The annual budget for the National Institutes of Health--supported by taxation--was over a billion dollars even before funds for cancer research and cure were included. A vast sum of private resources is invested in the same effort; no reasonable person would object. But why do government and philanthropic dollars flow toward a cure for individual body cancer so copiously, and so meagerly or not at all toward a cure for the most horrendous disease in the body of society, namely war?

Money power is of course a material matter, but the motivations for its release are psychological or attitudinal considerations--not only the general attitude of the individual who releases the financial energy but his perspective about the institutions to which it is released. No institution of which we are aware operates without moneypower; we have too long been without a blueprint for peace research funding.

REFERENCES

[1]This information is from the *Statistical Abstract of the U.S. (1971)*, published by the U.S. Department of Commerce, p. 305.

[2]From *World Military Expenditures 1966-67* (United States Arms Control and Disarmament Agency, 1968), p. 8.

[3]From *World Military Expenditures,* p. 8.

[4]From *World Military Expenditures,* p. 5-6.

[5]Lewis Mumford is quoted from *The Myth of the Machine* (New York: Harcourt, Brace and World, Inc., 1966), p. 281.

[6]This information can be found in *The Foundation Directory, Edition 4,* edited by Marianna O. Lewis (New York: Columbia University Press, 1971), p. xi.

9. Summary Propositions

We have examined the distance between current peace effort and technological peace effort. No real appetite for the change has been whetted up till now; this should alter when there is a greater awareness of the weakness, not to say futility, of current peace effort and a clearer vision of what change technology might bring. We have approached the gap with two kinds of surveys--the causes of peace effort failure and the new methodology that would increase the chances of success.

In this final chapter we shall summarize the propositions made in previous chapters; it will be apparent that the summarizations are at once antecedents and could be presented, with equal logic, as an opening or closing statement.

1. As soon as possible the creative thinkers in this field need to develop an acute sensitivity to the relationship between staff size and task size, lest we continue indefinitely to misplace our available energies. We have assailed prematurely the total foreign policy of powerful nations, and made a direct attack--too early and too little--upon the industrial-military complex of the United

States and other nations similarly geared. At the same time, we have not sufficiently recognized the essential inertia of the masses nor developed antidotes, such as the re-education of educators, mothers or any group in the particular position of influencing ethical inclinations. In all these areas we should make careful observations to discover what limited success has been gained, at the same time identifying the factors which have set the limits.

In a random interview survey of opinion on peace movement failure, most responses were summable under "size of the *task*"--the respondents enumerating the various blockades to peace but rarely citing the smallness of the *task force.* The task force that does exist--current peace researchers--seems bent on the study of war and very little inclined to the study of peace, as pointed out by Werner Levi in an article in the *Journal of Conflict Resolution.*[1]

2. Since we are willing to admit to the shortcomings of physical technology, we need to be equally sensitive to the uneven progress in social matters. It is a poignant fact that neither racism nor poverty nor war have been eliminated. Not yet. But this must be our challenge and hope. The fringe negatives of physical technology are certainly there--we shall undoubtedly refine that technology rather than eliminate it. There must be a parallel development of new social and moral motivations, resulting in the democratization of all technology, whether physical or social.

The antipathy to technology in general and antagonism towards extending scientific method to social and human relations is nothing new to peace researchers. One French thinker-writer illustrates this point. This is

Jacques Ellul, whose book, *The Technological Society,*[2] prods us by its pessimistic and negative viewpoint. The effect of Ellul's book need not be negative, however, if we take care to make it provocatively useful.

In *Towards a Science of Peace,* the position was presented that it is not logical to abandon the proven power of scientific method, but that the hope lies in developing a new direction for the application of the scientific attitude. So one slogan that might grow out of this present book is that "the cure for bad technology is best found in new and better technology"--technology in new directions.

3. Apparently there are only a few studies which implicitly or explicitly appear to be directed by technological orientation. There is very little research concerned with the problem of manpower for peace. Only one good study has come to our attention which examines the problem of moneypower--see Alcock and Laulicht in the bibliography.[3] Such crucial attitudinal dimensions as belief in the eliminability of the institution of war seem to have been almost completely overlooked. An inspiring exception is David Krieger's research at San Francisco State College, California, U.S.A.

Careful, concerned observers feel strongly the need for strengthening the United Nations, although there is not perfect agreement as to the immediate direction of the change. Very little research has dealt with the imperative of improving the financial structure. Withey and Scott have done an excellent job in their effort to summarize the attitudinal dimensions of the problem. Drawing from public opinion polls in the United States, they have traced attitude fluctuations towards this organization over a ten year period. There is a factor of inadequacy in their data, inasmuch as their opinion poll data was drawn from items interjected by organizations primarily concerned with market research. A detailed view of their work can be had by reading a review of their summarization by Lentz, in the *Journal of Conflict Resolution.*[4]

4. We must remind ourselves repeatedly that research is not only for knowledge, but for applicable knowledge and for the techniques of effective application. An appreciable amount of data has accumulated as peace research has struggled to get on its feet. The neglect of the criterion of applicability makes the technological approach imperative.

Who and where are the persons already equipped with the essential attitude, antecedent to the development of a technology of peace? The need for an appropriate tool for finding them is evident. The St. Louis Peace Research Laboratory has made a start on an Opinionaire designed to discover and measure this attitude among potential workers. One assumes these pioneers will attract counterparts. This Peace Technology Attitude Opinionaire is designed not only to seek out those with a certain technological attitude, but also those who exhibit a healthy discontent with our lack of progress in attaining peace.

The pencil and paper survey with PTAO could well be followed by more intensive personal interviews; the process could begin without delay in existing peace research centers, peace institutes, and among students in the proliferating courses for the study of peace in colleges and universities. The use of the Opinionaire would begin within a community, but would expand systematically until its thrust is world-wide. At each expansion, adequate time should be given to the study of the findings. This will include both correlational study and validity judgement. The latter will be based not only on the effectiveness of the selection process, but also the Opinionaire's usefulness as an initial means of intercommunication among those whom it identifies.

5. *If the assumption is correct that manpower and moneypower lie at the top of the priority list of problems, then severe handicap lies in uncertainty regarding the extent or supply of risk capital, both financial and intellectual.* So it is hoped that enough individuals may be found who are willing to lead off and not wait for others to act first; individuals so sensitive to the need that they are willing to risk their own capital of time and money and prestige in order to get started.

There are further reminders to be made. We tend to speak in terms of the potential help from the moral and existing peace elite. Even though the potential is not high, we should not ignore the power elite--including the financially powerful. Why not count on the wealthy as well as the not so wealthy? Someone has observed that peace researchers are on the "left" and the wealthy are on the "right," however exceptions do exist and should not be overlooked.

The fact that there are deviations to the pattern makes a case for specificity. As in the study of other attitudes, our assumptions concerning the alignments of wealth and peace attitude have been predicated on very incomplete data. It has been found in the measurement of conservatism-radicalism by the St. Louis Peace Research Laboratory that the basis for this specificity doctrine has disappeared with the multiple item and broad measurement tool. Now, however, the tool needs further broadening and refinement to take care of the exceptions.

6. This leads to another caveat for those who are interested in the developmental approach--beware of the danger of undue premature parsimony. Public opinion polls are invariably oversimplified and their brevity should be imitated as little as possible. Measurement tools should be flexible and evolutionary. At this point, our Peace Technological Attitude Opinionaire has fifty items. In the developmental stages, it will probably be extended greatly; later, on the basis of the findings, correlational and otherwise, it may be reduced back to original length--perhaps even less. For broad exploratory purposes the briefer opinionaire has distinct advantages, but effectiveness need not be sacrificed to brevity.

7. Current peace effort seems to take piecemeal form; we must neither be misled by fragmentation, nor neglectful of due consideration--some fragments need merely to be amplified and integrated. It will be useless to buffet such mountains as mass inertia or the military-industrial juggernaut without appropriate force--better to clear the trails leading to a more adequate task force and then to more effective tools. The jet-engined plane is a better mechanism than the sailing vessel, but the shift did not come overnight. The first plane carried one person for seconds--now our planes carry hundreds and stay aloft for many hours. We expect to produce the eventual mechanisms for transcending the barriers between us and a warless world, and we seek to achieve this with a technology which we have already defined as a combination of mechanisms, techniques and tools for getting us from where we are to where we want to go.

8. One of our watchwords must be "contact." There is no way to overemphasize the value of communication.

Relevant conversation with self and with others is greatly needed for momentum. By intercommunication we discover not only what is best within the thinking of our contemporaries, but what is good within ourselves, as well. Dialogue, in its best form—probing, vigorous, but modest and not dogmatic—exposes erstwhile naivety, introduces hypotheses, and thus stimulates productive interaction.

9. At no point can we neglect morale, however much patience this requires! As more sparks fly from the imaginative anvil and more persons are attracted to peace technology pioneering, the stronger their morale--a welcome prospect since it might well be said that thus far our Achilles' heel has been our lack of conviction--war has us "licked" because we think we are licked. There are some guidelines for the sort of interaction that will reinforce rather than undercut morale:

Dogmatic controversy, critical, negative and vindictive, is far less productive than a consistent, openminded focus on one's own opportunities and fundamental convictions.

Much energy is wasted in the fruitless attempt to change the attitudes of others; attempt, instead, to discover and utilize appropriate attitudes wherever they are found.

Fragmented peace efforts and distractive peace competitives must be amplified and integrated so that the individual may orient himself according to his interests and capabilities.

Sporadic, excoriating rhetoric--no matter how sincere--will accomplish little compared to the investment of one's energies in the structuring of a humane social and political milieu.

10. There is a growing movement towards research on the future. Most persons so engaged are extrapolators dealing only with what is probable; only a few are concerned with what the future could or should be. *More and more, desirability should become, if not a substitute, at least of equal weight with the concern for probability.* Prediction need not be abandoned as one of the touchstones of science but needs to be improved by technological orientation--not merely what is likely to happen by the year 2000, but what could and ought to happen. As two of our colleagues in Britain have said, "Let us strive to make the most desirable the most probable."[5]

11. Success in technology for peace as well as in a vast array of other social and political futures does not have to be attained at a single jump. There are instructive parallels with some of the factors making for success in past political revolutions. One of these has to do with the dynamics of a small but effective minority at the onset. Only gradually do the masses come into line. Always the crude processes of bringing about changes have caused injustices--too often, great bloodshed. Had there been available a social technology for political problem solving, some of these gruesome results might have been avoided. Some contend that war should be retained at least as an instrument of liberation. The goal of peace science and technology is to find a substitute for war which will not only do less violence to our moral sensibilities but also be more likely to prove effective and permanent in the maintenance of human life.

REFERENCES

[1]The article referred to is by Werner Levi, "On the Causes of Peace," *The Journal of Conflict Resolution,* VIII, 1 (March 1964), pp. 23-35.

[2]Originally published in France, the translation referred to is Jacques Ellul, *The Technological Society* (New York: Alfred A. Knopf, 1965).

[3]The article by Jerome Laulicht and Norman Z. Alcock is "The Support of Peace Research," *The Journal of Conflict Resolution,* X, 2 (June 1966), pp. 198-208.

[4]See Theo. F. Lentz, "Public Opinion Research for Peace: A Review and Discussion of William A. Scott and Stephen B. Withey, The United States and the United Nations: The Public View, 1945-1955," *The Journal of Conflict Resolution,* IV, 2 (June 1960), pp. 234-242.

[5]The concept comes from Stafford Beer, "The World, the flesh and the metal," *Nature,* 205, 4968 (January 16, 1965), pp. 223-231 and is developed by Paul Smoker in *Simulation for Social Anticipation and Creation* in research for the Advanced Research Project Agency at Northwestern University, Evanston, Illinois, U.S.A. in 1969, p. 26.

SELECTED BIBLIOGRAPHY

Beer, Stafford. "The World, the flesh and the metal." *Nature,* 205, 4968, January 16, 1965, pp. 223-231.

Benoit, Emile. "The Economic Impacts of Disarmament." *Behavioral Science and Human Survival,* ed. Milton Schwebel. Palo Alto, California: Science and Behavior Books, Inc., 1965.

_____, and Kenneth Boulding, eds., *Disarmament and the Economy.* New York: Harper and Row, 1963.

Boguslaw, Robert. *The New Utopians.* Englewood Cliffs, New Jersey: Prentice-Hall Inc., 1965.

Eckhardt, William and Alan G. Newcombe. "Militarism, Personality, and Other Social Attitudes." *The Journal of Conflict Resolution,* XIII, 2, June 1969, pp. 210-219.

"Ecology Manifesto: The Power to Create." *Our Generation,* VII, 1, March 1970, p. 99.

Ellul, Jacques. *The Technological Society.* New York: Alfred A. Knopf, 1965.

Erlich, Paul R. and John P. Holdren. "Technology for the Poor." *Saturday Review,* LIV, July 3, 1971, pp. 46-47.

The Foundation Directory, Edition 4, ed. Marianna O. Lewis. New York: Columbia University Press, 1971.

The Futurist. Washington, D.C.: World Future Society: An Association for the Study of Alternative Futures.

Garcia, John David. *The Moral Society a rational alternative to death.* New York: The Julian Press, Inc., 1971.

Hughes, Langston. "Harlem." *Montage of a Dream Deferred.* New York: Henry Holt and Company, 1951.

International Repertory of Institutions Specializing in Research on Peace and Disarmament, eds. Johan Galtung and Marie Rugi. UNESCO, 1967.

Journal of Peace Research, ed. Johan Galtung (International Peace Research Institute, Oslo, Norway). Groningen, Netherlands: International Peace Research Association.

Laulicht, Jerome and Norman Z. Alcock. "The Support of Peace Research." *The Journal of Conflict Resolution,* X, 2, June 1966, pp. 198-208.

Lentz, Theo. F. "Public Opinion Research for Peace: A Review and Discussion of William A. Scott and Stephen B. Withey, The United States and the United Nations: The Public View, 1945-

1955." *The Journal of Conflict Resolution,* IV, 2, June 1960, pp. 234-242.

———. *Towards a Science of Peace.* London: Halcyon Press (Barnet) Ltd., 1955.

Levi, Werner. "On the Causes of Peace." *The Journal of Conflict Resolution,* VIII, 1, March 1964, pp. 23-35.

Mankind 2000, eds. Robert Jungk and Johan Galtung (International Peace Research Institute, Oslo, Norway). Oslo Universitetsforlaget: London, Allen and Unwin, 1969.

Mumford, Lewis. *The Myth of the Machine.* New York: Harcourt, Brace and World, Inc., 1966.

———. *Technics and Civilization.* New York: Harcourt, Brace and Company, 1934.

"Peace Career Expressionaire." (Peace Research Laboratory, St. Louis, Missouri, U.S.A.).

Peace Research Abstracts Journal, eds. Alan and Hanna Newcombe. Oakville, Ontario: Canadian Peace Research Institute.

"Peace Technological Attitude Opinionaire." (Peace Research Laboratory, St. Louis, Missouri, U.S.A.).

Rapoport, Anatol. "Can Peace Research be Applied?" *The Journal of Conflict Resolution,* XIV, 2, June 1970, pp. 277-286.

Skinner, B.F. *Beyond Freedom and Dignity.* New York: Alfred A. Knopf, 1971.

Smoker, Paul. *Simulation for Social Anticipation and Creation.* (Advanced Research Project Agency). Evanston, Illinois: Northwestern University, 1969.

Statistical Abstract of the U.S. (1971). U.S. Department of Commerce.

Toffler, Alvin. *Future Shock.* New York: Random House, 1970.

World Military Expenditures 1966-67. U.S. Arms Control and Disarmament Agency, 1968.

INVITATION TO THE READER

Now that you have read our brief presentation perhaps you would like to cooperate with a response of some sort. As stated in the preface, our hope has been to present a challenge that will evoke dialogue.

It has taken much audacity to write this book. We have been tempted to wait until we had time to gain a better perspective through consultation with individuals whose backgrounds are better than or at least different from ours. However, time is of the essence. It may be now or never to launch a peace technology movement. Such a movement will take time for growth.

We should offer more illustrations of peace technology, if not factual then imaginative, if not specifically peace related then of social technology in general.

Let us have your criticisms of technology in a broad sense and any fears you have of technology when applied directly to the behavior of persons--how can we avoid undemocratic manipulations of human individuals?

In 1965 Norman Cousins stated, "The biggest lesson of all to be learned about contemporary civilization is that nothing anyone is doing today makes any sense unless it is connected to the making of a genuine peace."

Perhaps we have not made clear the distinction between science and technology--can you point out our gaps? Perhaps you would tell us how you would like to participate--with moral support or intellectual or financial.

Let us take advantage of whatever time remains--of our chances to achieve peace in time to achieve human survival.

Theodore F. Lentz, Director
Peace Research Laboratory
6251 San Bonita
St. Louis, Missouri 63105, U.S.A.